10 SECONDS TO

C000192732

10 Seconds

To

Child Genius

TRAY-SEAN BEN SALMI

"I'm That Kid"

(Channel 4 TV Child Genius)

and

PHILIP CHAN

"10-Seconds Maths Expert"

Awards Winning Authors

ISBN: 099286948X
ISBN-13: 978-0992869489

DEDICATION

THE CO-AUTHORS (TRAY-SEAN AND PHILIP)

We want to dedicate this book to

YOU our READER

because we know
**that deep down inside
YOU is a Genius**
just waiting to come out and play.

The fact you have purchased this book means you are not a Passive person waiting for things to happen but you are a **DOER and ACTION TAKER,** as one of the important traits of any successful Man or Woman.

From TRAY-SEAN BEN SALMI

I would like to dedicate this book to:
My Step-Father and Mother,
Mohamed and Sabrina Ben Salmi.

My older Sister **Lashai Ben Salmi,**
Aka My Journey,

My Grandmother **Mary Paul (and Marcia Phinn)**
Aka Let go and live,

My Friend from 4 years old **Shane Gordon**

Sunil Chutni

Mich'el Pierre (Professional Singing Artist)

Yasmine Ben Salmi,
Aka Love Prenure,

Paolo Ben Salmi,
Aka Pint Size Adventurer

and

My youngest brother
Amire Ben Salmi,
Aka Mr. Because I AM Intelligent.

From PHILIP CHAN

I want to dedicate this to a few other very Special people, friends that have supported me through thick and thin:

John and Beverley Neilson, life long friends.

Terry Andrews, one of the most inspiring Mathematics Teacher in the UK.

Mayooran Senthimani, for his technical knowledge to make these publications possible. He is also a prolific writer and an Award Winning Author in his own right.

Lastly, the driving reasons why I am working beyond my retirement years as my legacy to nine very Special young people in my life and whom I love dearly:

Joshua & Fynn Whiting, Kristie and Ben Welch & Freya Whiting, Leah, Moya, Chloe and Jack Whiting

I hope when they grow up, they too will find their talents and strengths to Make a Difference in the World.

WHY DID WE WRITE THIS BOOK?

TRAY-SEAN' STORY

At school, I was a very shy kid and so lacking in confidence.
I wanted to change for the better and together with my eldest sister, Lashai, we started to learn about Personal Development from some of the top people in the UK and the world, with wonderful encouragements from our mother Sabrina Ben Salmi.

Over a period of time, we have met so many incredible people, coaches and trainers who help us to develop our confidences and started to believe in ourselves.

Our Millionaire mentors often tell us

"If you want to succeed in Life,
then get comfortable being Uncomfortable and
get out of your comfort zone"

Wow, this is hard stuff but we trusted them and starting doing the things they advised us to do.

So I decided to get out of my comfort zone and applied to enter Channel 4 TV 'Child Genius' competition competing against the best 8 – 12 year olds in the UK and I managed to pass all their tests and made it to the Child Genius' 2017 Finals as one of the Top 20 contestants in the country.

I am over the moon, "Little shy ME, made it to the Final!"

The reason I am writing this book is because someone believed in ME (my lovely Mother) then other great mentors and as a consequence I am starting to achieve things that was once beyond my wildest dreams.

I want to BELIEVE IN YOU and to tell you how
amazing you are and in this book:
"10 Seconds to Child Genius"
with my friend and mentor (and co-author),
we want to encourage YOU to make your Dream
come true.

YOUR DREAM!

You have to understand F.E.A.R. is really
False Evidence Appearing Real

But I want YOU to understand another thing about
F.E.A.R.
And that is when you face FEAR in the face.

F.E.A.R. – Face Everything And Rise.

Yes Friend, I would love to be your Friend and say
again:

I BELIEVE IN YOU!

PHILIP'S STORY

"Go away you stupid boy"

These were the haunting words from over 60 years ago when I was at school that made me felt a shamed, insignificant and useless!
I worked hard but still I was a poor performing student and low achiever.

YOU SEE WORKING HARD IS NOT THE ANSWER – IT IS ABOUT WORKING SMART THEN WORKING HARD IS THE KEY

But as a young boy, I didn't know any better.

My saving grace was an old teacher, Mr. Roland Duff and he said three Magic sentences that started to transform my life!

LCS 1 (Life Changing Statement)

**"There is no such thing as a stupid question if you honestly don't know
– so learn to ask quality questions!"**

LCS 2

**"It is better to FEEL stupid for 5 minutes then to STAY STUPID for the rest of your life
– so learn to ask for help regularly!"**

LCS 3 (The most Life Changing Statement for me!)

**"I BELIEVE IN YOU
– if you are serious, I will help you!"**

BECAUSE someone, in this case, Mr. Roland Duff believed in me, he helped me to changed my MINDSET from getting

<u>NOW</u>HERE to **NOW HERE**
as 'The 10 Seconds Maths Expert'

Both Tray-Sean and I,

"WE BELIEVE IN YOU"

Let us say it again, LOUDLY:

"WE BELIEVE IN YOU"

Go after <u>YOUR GOALS</u> and <u>DREAMS.</u>

Go and discover your hidden talents and <u>Your Uniqueness</u>.

Keep on keeping on until you reach your goals and dreams.

We (Tray-Sean and I) would love to be your Mentor and Coach.

You can contact us on:

www.10secondstochildgenius.co.uk

Facebook: 10 Seconds To Child Genius

CONTENTS

ACKNOWLEDGMENTS

We have been so grateful to a number of very inspirational people
who have coached, mentored or greatly influenced us
either directly or indirectly.
They are some of the world's finest people.
All of them are Number One in their respective fields of expertise.
Superb coaches and Leaders like:

**JT Foxx, Gerry Robert,
Andy Harrington,
T Harv Ecker, Nick Vujicic,
Jack Welch, George Ross,
Shaa Wasmund,
Lady (Michelle) Mone,
Richard Branson,
Steve Wozniak, Jay Abraham,
John Assaraf, Bob Proctor,
Tony Robbins,
Dr Nido Qubein and the late
Jim Rohn.**

**Shamilla Soosaipillai,
Giselle Malawer, Sammy Blindell,
Miles Fryer, Tosin Ogunnusi.**

Dr Cheryl Chapman,
Dr Marina Nani, Dr Darie Nani,
Kris Kemery Toone, Robert G. Allen
& Angii Anderton, Debbie Stoute,
Ramasamy Kavitha.

Oprah Winfrey,
Stedman Graham,Gordana Biernat,
Neale Donald Walsch,
Dr John Demartini,
T. Harv Eker and
Robert T. Kiyosaki.

Melanie Stewart. Junior & Lauren
Ogunyemi. Tosin & Anna Ogunnusi,
Michelle Watson, Elaine Summers &
Jade. Majid Khan.
White Out Originals:
Mahmood Shaikh & Team.

Brown Rudnick:
Georgie Collins & Team,
Denise Ramsey & Unltd.

Child Genius Team and my peers
and their families.

Our tutor and spiritual grandfather
Ras Al.
Lorna Blackman and her
phenomenal theatre school and many
more

I would also like to take this opportunity to acknowledge those of
you who delivered us out of adversity in our lives.
This made us stronger and inspired us to convert our adversities
into empowerment for ourselves and others.
After all, as a direct result of everything we experienced, this has
lead us to changes and growth in us beyond our wildest dreams.

"Every adversity, every heartache carries within it the Seeds of an Equivalent or Greater Benefit"

We always want to constantly learn from the very best people. Recently we are very fortunate to have the coaching and mentoring of great International business mentors:

Douglas Vermeeren, Regan Anne Hillyer and Juan Pablo Barahona.

"Every great performer some time, somewhere NEEDS a coach and a team to support him or her."

In fact, the richest and most successful people we know, some have 2 to 8 coaches or mentors in different areas of their life!

In reality, THERE IS NO SUCH THINGS AS A SELF-MADE MILLIONAIRE!

To succeed in ANYTHING, you need a team of people who can do the things you are not good at, so you can focus on YOUR STRENGTHS and the things YOU LOVE TO DO!

For Tray-Sean and I, we have a fabulous team:-

1. **The Ben Salami Family member and our darling Grandma – Mary Paul.**

2. **Mayooran Senthilmani – our Publisher.**

3. **John Neilson – For editing of our book.**

4. **Prasanthika Mihirani – Our Creative Book Designer.**

5. **Donna Hudson – Our IT expert for her help in creating our website.**

6. **Emma Hollings – The Winning Award professional photographer helping to promote our events**

7. **Chloe Whiting – my 10 year old Grand daughter for giving an old retired person a reason to get off his rocking chair!**

Here are some of the people and Institutions that already have a copy of my book.

Her Majesty Queen Elizabeth II

Rt. Hon. David Cameron (Former British Prime Minister)

British National Library

Bodleian Libraries (University of Oxford)

Cambridge University Library

Raymond Aaron
(New York Times #1 Bestselling Author)

J T Foxx
(World's #1 Wealth and Business Coach)

Andy Harrington
(The World's Leading Public Speaking Coach)

Gerry Robert
(Owner of Black Card Books and International Bestselling Author)

Author's notes

For Readers in USA and Canada

The abbreviations for 'Mathematics' in UK is:
' Maths ' instead of ' Math ' in USA and Canada

Published by

DVG STAR Publishing

Regarding the wonderful designs for our book cover,
We must give credit to our amazing Creative Designer,
PRASANTHIKA MIHIRANI.

You can find her on Facebook as

Swiss Graphics

Ten Things You Should Know About Your Brain

FACT ONE:
There are always four Stages of Learning

Stage One - UNCONSCIOUS INCOMPETENCE

This is when you start doing something new by 'Trial and Error' and hoping it will work. Does it? Hmm!

Stage Two - CONSCIOUS INCOMPETENCE

This is the 'MAKE or BREAK' stage and often 95% people will give up at this stage if they don't get success quickly and as a consequence, they will NEVER discover their real potential! So it is vital to get a Mentor, Coach or an Instructor.

For example, if you want to learn to drive a car then book a series of lessons with a Qualified Instructor.

Stage Three - CONSCIOUS COMPETENCE

Yeah! You can do it!
Like after a number of lessons you know how to drive a car but you still need to think of everything you are doing to drive the car safely!

Stage Four - UNCONSCIOUS COMPETENCE

You have mastered the skill because you had so much practice and can do it without thinking about it.

FACT TWO:
We all have a Preferred Learning Style known as V.A.K.

V – VISUAL (We like to learn by Reading or Looking at the information);

A – AUDITORY (We like to learn by Listening or Talking);

K – KINESTHETICS (We like to learn by Doing and Active).

To increase our skills factor, ideally we want to develop over time all three V, A and K!

FACT THREE:
We actually have 5 Types of W.I.R.E.S. Memory

W – WORKING MEMORY (Short Term Memory)

I – IMPLICIT MEMORY (Or sometimes called 'The Muscle' memory. Once you have learned to do something, like how to use a new computer software etc)

R – REMOTE (This is your lifetime accumulation of skills and knowledge and seems to diminish with age if you don't use it! "USE IT or LOST IT!")

E – EPISODIC (This is when you have a memory of a specific experience or an event.

S- SEMNATIC (This is when certain words and symbols are

special to the individual).

FACT FOUR:
Learn to Search and Recognized for Patterns enhances Brain development.

FACT FIVE:
Your ability to learn is 'State' dependent So have a High Expectation when you learn you will succeed.

FACT SIX:
Emotions and Learning are closely linked.

Watch what you are saying to yourself when learning.
Have a high expectation of yourself.

Don't say:
"I will never be able to learn all these things!"

Instead, say something like:

"Everything I learn I will remember at the right time to use."

FACT SEVEN:
We all have TWO Daily Learning cycles

The two cycles are: "Low to high energy" and "Relaxation to Tension" Cycle.
So be aware of your best time for learning, especially for test revision.

FACT EIGHT:

Our brain modal switch over roughly every 90 minutes.

Generally speaking, our left brain is more efficient for verbal skills and our right brain for spatial skills.

FACT NINE:

Our Learning and Physical performance is affected by our biological rhythms throughout the day.

Even our breathing has cycles.
Overall, short term memory is best in the morning and not so effective in the afternoon.
Whilst our long term memory is better in the afternoon.

FACT TEN:

Your brain needs Deep Relaxing sleep.

This allows time for your brain to process all the things you have learned.
Getting into a REM sleep (Dreaming) has been found by researchers to be very important for learning.

CHAPTER ONE

THE OBVIOUS IS NOT OBVIOUS!

Here is a famous Optical illusion.

What do you see?

For some people, they will see it straight away but for others they will need to take a bit of time!

Can you see what the 3 letters word?

If you still can't see it, then focus your eyes in the blank space rather than the black area.

Can you see it now?

Once you have SEEN the word, you cannot NOT see it!

Right!

Yes, it is the word 'F L Y'

Now close your eyes and open it again.

Now it is obvious. Isn't it!

"The obvious is not obvious until it is obvious"

Does that make sense; Yes or No?

Let me explain what I mean with a few more examples.

What is 3 x 4?   3 x 4 equals 12   or   12 = 3 x 4

That is: 1 ....2....3.....4

Should I say that again, it is 1....2....3....4....
(Obvious –isn't it?)

What is 7 x 8?   It is 56 - So 7 x 8 is 56...and 56 = 7 x 8

So it is 5...6...7...8! .... So it is 5...6...7...8

Can you see the obvious **IS** now obvious?

Personal Notes

Not all times tables follow a pattern, so sometimes we may have to use another technique to help us.

We all remember silly things, so say this after me:

3 x 7 is 21 because mum and dad always look 21

Just for fun work tonight, you have to say this 100 times before falling asleep (Joke!)

Here is another one just for fun, say aloud:

8 times 8 is 64 because I ate (8) and ate (8) and

I was Sick (6) on the floor (4)

So 8 eights is 64   that is   8 x 8 = 64

Now say this out loud three times

Louder please.

"*I ate and ate and was sick on the floor, so eight eights is 64*"

You can make up some yourself with a table that you find difficult to remember. Now go and have some fun with it!

We remember silly things, so you have my permission to be 'silly' with your times table.

CHAPTER TWO

THE QUICK '8'

When you learn to spot patterns,
then you can

TSBD – The Same But Different!

Let us share with you another idea to learn your 8 times
tables.

Here is the normal 8 times table:

$$8 \times 1 = 08$$
$$8 \times 2 = 16$$
$$8 \times 3 = 24$$
$$8 \times 4 = 32$$
$$8 \times 5 = 40$$
$$8 \times 6 = 48$$
$$8 \times 7 = 56$$
$$8 \times 8 = 64$$
$$8 \times 9 = 72$$
$$8 \times 10 = 80$$

What did you notice about the last column (the UNIT
DIGITS?)

If you have seen the last digits are:

8-6-4-2-0 and it repeats 8-6-4-2-0 if you extend the 8 times table. Well done!

This gives us a key how to learn the 8 times table another way. This is what we want you to do:

Write, in first column   : 8 7 6 5 4 4 3 2 1 0
Write, in the Second column : 0 2 4 6 8 0 2 4 6 8

It should look like this:

FIRST COLUMN	SECOND COLUMN
8	0
7	2
6	4
5	6
4	8
4	0
3	2
2	4
1	6
0	8

The obvious is not obvious until it's obvious!

Let me explain what we mean by this.

If you look carefully, this is the 8 times table in reverse:

$$80 = 8 \times 10$$
$$72 = 8 \times 9$$
$$64 = 8 \times 8$$
$$56 = 8 \times 7$$
$$48 = 8 \times 6$$
$$40 = 8 \times 5$$
$$32 = 8 \times 4$$
$$24 = 8 \times 3$$
$$16 = 8 \times 2$$
$$08 = 8 \times 1$$

So by writing in a column:

8   7   6   5 | 4   4 | 3   2   1   0

Be careful of the Two 4's in the middle.
The simple way to remember this :
Two 4's make 8

Then second column:

0 2 4 6 8   0 2 4 6 8

You can now do practise the 8 times table quickly !

Enjoy and have fun

Personal Notes

CHAPTER THREE

SQUARING NUMBERS ENDING IN '5'

What is 5 x 5?
It is of course 25 simple, right?

Let's see how we can use this simple information to help us work out square numbers that ends in '5'.

For example:

$$5^2 = \ 2\,5^2 = 2\,\underline{5} \times 2\,\underline{5}$$
$$= 6\,\underline{25}$$

Let's see how we can do this quickly.
Sometimes when to take time to save time, this is an important skill.

When you multiply a 2-digit square number that ends in '5', The last TWO digits will always be 25.

Step One: 2 $\underline{5}^2$

Multiply the unit '$\underline{5}$' by itself: $\underline{5}$ x $\underline{5}$ = 25

Step Two:

There 2 methods to deal with the Tens digit '2'

Method 1: (2 + 2) + 2 = **6, or**

Method 2: 2 x (2 + 1)
$\qquad\qquad$ = 2x3 = **6**
So 25^2 = **6** $\underline{25}$

Here is another example: 75^2
 Step 1: 7 $\underline{5}^2$ = _ _ 25

Step 2: (7 + 7) + 7 =56
or
7 x (7+1) = 7 x 8 = 56
So, 75^2 = $\underline{56}$ $\underline{25}$ = 5625

Now it is your turn to complete this exercise then go and amaze your friends.

(You may want to time yourself)

$15^2 =$

$25^2 =$

$35^2 =$

$45^2 =$

$55^2 =$

$65^2 =$

$75^2 =$

$85^2 =$

$95^2 =$

(Answers on next page)

How did you get on?

Here are the solutions:

$\underline{1} \, 5^2 = \underline{(1 \times 2)} \quad 25 = \quad \underline{2} \, 25$

$\underline{2} \, 5^2 = \underline{(2 \times 3)} \quad 25 = \quad \underline{6} \, 25$

$\underline{3} \, 5^2 = \underline{(3 \times 4)} \quad 25 = \underline{12} \, 25$

$\underline{4} \, 5^2 = \underline{(4 \times 5)} \quad 25 = \underline{20} \, 25$

$\underline{5} \, 5^2 = \underline{(5 \times 6)} \quad 25 = \underline{30} \, 25$

$\underline{6} \, 5^2 = \underline{(6 \times 7)} \quad 25 = \underline{42} \, 25$

$\underline{7} \, 5^2 = \underline{(7 \times 8)} \quad 25 = \underline{56} \, 25$

$\underline{8} \, 5^2 = \underline{(8 \times 9)} \quad 25 = \underline{72} \, 25$

$\underline{9} \, 5^2 = \underline{(9 \times 10)} \, 25 = \underline{90} \, 25$

Now go and impress your friends!
Teach them how to do it as well.

Personal Notes

CHAPTER FOUR

SQUARING NUMBERS BEGINNING WITH '5'

What happens if the number begins with a '5' for a 2 digit square number ?

Rather than simply telling you how to do it,
I want to see if you can work it out for yourself ?

However, I will give you a few clues to guide you.

For example

53^2 means 53 x 53, which equals 28 09

You should be able to do this in less than 5 seconds.

Here are the hints:

Hint One:

What did we do to '5' to get 28?

Hint Two:

What did we do to '3' to get 09?

Solution on next page.

Check to see if you have worked it out!

$53^2 =$ **5** $\underline{3}$ x **5** $\underline{3} =$ **26** $\underline{09}$

Step One: **26 = (5 x 5) +** $\underline{3}$
Step Two: $\underline{3}$ x $\underline{3}$ = $\underline{09}$

So 53^2 = **28**$\underline{09}$

Try this one: **5** $\underline{7}^2$

Step one: **(5 x 5)** + $\underline{7}$ = 25 + 7 = 32
Step two: $\underline{7}$ x $\underline{7}$ = $\underline{49}$

So, **5**$\underline{7}^2$ = **32**$\underline{49}$

Now you can complete the table below quickly and easily!

$51^2 =$

$52^2 =$

$53^2 =$

$54^2 =$

$55^2 =$

$56^2 =$

$57^2 =$

$58^2 =$

$59^2 =$

(Answers on next page)

The Solutions

$51^2 = (5x5)+1 , 1x1 = 26\ 01$
$52^2 = (5x5)+2 , 2x2 = 27\ 04$
$53^2 = (5x5)+3 , 3x3 = 28\ 09$
$54^2 = (5x5)+4 , 4x4 = 29\ 16$
$55^2 = (5x5)+5 , 5x5 = 30\ 25$
$56^2 = (5x5)+6 , 6x6 = 31\ 36$
$57^2 = (5x5)+7 , 7x7 = 32\ 49$
$58^2 = (5x5)+8 , 8x8 = 33\ 64$
$59^2 = (5x5)+9 , 9x9 = 34\ 81$

Look at the answers.

Can you see a pattern?

The best way to remember new things is to show it to as many people as possible, as soon as possible!

Personal Notes

CHAPTER FIVE

PATTERNS IN SQUARE NUMBERS

This is the chart of the first 30 square numbers.

Chart of Perfect Squares 1 to 30

1^2	=	1	11^2	=	121	21^2	=	441
2^2	=	4	12^2	=	144	22^2	=	484
3^2	=	9	13^2	=	169	23^2	=	529
4^2	=	16	14^2	=	196	24^2	=	576
5^2	=	25	15^2	=	225	25^2	=	625
6^2	=	36	16^2	=	256	26^2	=	676
7^2	=	49	17^2	=	289	27^2	=	729
8^2	=	64	18^2	=	324	28^2	=	784
9^2	=	81	19^2	=	361	29^2	=	841
10^2	=	100	20^2	=	400	30^2	=	900

Can you see the patterns in Square numbers?

What is GENIUS?

Sometimes, it is seeing the things which others are not seeing when it is right in front of their eyes!

Let's learn to look at things differently and focus on the first 10 square numbers

$$1^2 = 0 \quad 1$$
$$2^2 = 0 \quad 4$$
$$3^2 = 0 \quad 9$$
$$4^2 = 1 \quad 6$$

$$\boxed{5^2 = 2 \quad 5}$$

$$6^2 = 3 \quad 6$$
$$7^2 = 4 \quad 9$$
$$8^2 = 6 \quad 4$$
$$9^2 = 8 \quad 1$$

$$\boxed{10^2 = 10 \quad 0}$$

Can you see the pattern
Look carefully at the 'UNIT' digits!

If you carefully, the pattern is:

1 4 9 6	5	**6 9 4 1**	0

Your Investigation

Either write out the first 1^2 to 100^2

Or print off a sheet of the first 1^2 to 100^2

Check to see if this pattern recurs and
Can you see other patterns as well?

Have fun and investigate!

We would love to hear from you any other patterns you have found.

Sometimes :

"The obvious is NOT obvious until it's obvious"

You may see patterns that no one else has seen!

Investigation Notes

Investigation Notes

Investigations Notes

CHAPTER SIX

QUICK '12'– PART ONE

THE SAME BUT DIFFERENT!
Here is another and cool way to do your 12 Times Table which we called the Quick '12' method.

The Power of Doubling

The key exercise (If you can do this 12 times table then multiplying by 12 it is easy!)

So just double the numbers and put the answer underneath in the box below:

1	2	3	4	5	6	7	8	9	10
11	12	13	14	15	16	17	18	19	20

12 times table	What is double	Answer
1 x 12	1 (double 1 is 2)	12
2 x 12	2 (double 2 is 4)	24
3 x 12	3 (double 3 is 6)	36
4 x 12	4 (double 4 is 8)	48
5 x 12	5(double 5 is 10, so add 1 to 5 makes 6)	60
6 x 12	6(double 6 is 12, so add 1 to 6 makes 7)	72
7 x 12	7(double 7 is 14, so add 1 to 7 makes 8)	84
8 x 12	8(double 8 is 16, so add 1 to 8 makes 9)	96
9 x 12	9(double 9 is 18, so add 1 to 9 makes 10)	108
10 x 12	10(double 10 is 20, so add 2 to 10 makes 12)	120
11 x 12	11(double 11 is 22, so add 2 to 11 makes 13)	132
12 x 12	12(double 12 is 24, so add 2 to 12 makes 14)	144

Personal Notes

CHAPTER SEVEN

QUICK '12' – PART TWO

Quick 12 -How to multiply any number by 12 quickly

Key skills:

1) To be able to double the number you want by 12

2) And then to add !

It's that simple !

A quick warm up ! (Use a calculator to check your answers)

Double these numbers :

NUMBER	DOUBLE THE NUMBER	NUMBER	DOUBLE THE NUMBER
5		47	
21		88	
33		67	
52		123	
63		312	
27		424	
38		612	

Question	Double this number	Write it like this	Put down the last digit at the end and add together the numbers underlined to get the final answer
14 x 12	14	*14* / *28* (14+2 = 16)	168
21 x 12	21	21 / 42 (21+4=25)	**252**
34 x 12	34	34 / 12 (34 + 1=35)	**352**
53 x 12	53	53 / 106 (53 +10 = 63)	**63 6**
67 x 12	67	67 / 134 (67 + 13 = 80)	804
123 x 12	123	123 / 246 (123+24 =147)	**1476**

Now it is your turn to have ago at these questions (you can check it with a calculator after you have done it on paper :

1) 23 x 12   2) 36 x 12   3) 48 x 12

4) 73 x 12   5) 113 x 12   6)   214 x 12

Remember:

24 x 12 is the same as 12 x 24,

You double the number that is multiplied by 12.

Personal Notes

CHAPTER EIGHT

RAINBOW MULTIPLCATIONS

This is a fun way to multiply two 2-digits numbers, which we call:

"THE RAINBOW MUTLPICATION"

Unlike normal multiplications, we work from RIGHT to LEFT, just like the way we would read a book.

In fact, many of the very top mathematicians in the world when they do this mentally, works from RIGHT to LEFT !

The 3 step Rainbow method:

STEP ONE: Multiply the 'Ten' digits together first
Then draw a line to the right;

STEP TWO: Now multiply the 'Unit' digits and finally;

STEP THREE: RAINBOW multiply and add!

Let's take you through step by step.

Example One: 12 x 13

$$\mathbf{1}\,2 \times \mathbf{1}\,3$$

Step One: 1 x 1 = 1

$$\mathbf{1}\,2 \times \mathbf{1}\,3 = \mathbf{1}\,\_$$

Step Two: 2 x 3 = 6

$$1\,\mathbf{2} \times 1\,\mathbf{3} = 1\,\_\mathbf{6}$$

Step Three: RAINBOW!

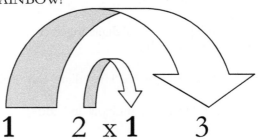

$$1 \quad 2 \times 1 \quad 3$$

$$
\begin{array}{rr}
1 \times 3 = & 3 \\
2 \times 1 = & +2 \\
\hline
& 5
\end{array}
$$

Therefore, 12x13 = 1<u>5</u>6

Example Two: 14 x 2 1

$$\mathbf{1}4 \times \mathbf{2} 1$$

Step One: 1 x 2 = 2

$$\mathbf{1}4 \times \mathbf{2} 1 = \mathbf{2}\_$$

Step Two: 4 x 1 = 4

$$1 \, \mathbf{4} \times 2 \, \mathbf{1} = 2 \_ \mathbf{4}$$

Step Three: RAINBOW!

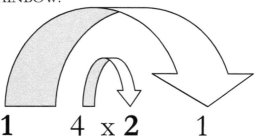

$$\mathbf{1} \qquad 4 \times \mathbf{2} \qquad 1$$

$$
\begin{array}{rr}
1 \times 1 = & 1 \\
4 \times 2 = & + \, 8 \\
\hline
 & 9
\end{array}
$$

Therefore, 14x21 = 2<u>9</u>4

Example Three: 34 x 5 2

$$\textbf{3}\ 4 \times \textbf{5}\ 2$$

Step One: 3 x 5 = 15

$$\textbf{3}\ 4 \times \textbf{5}\ 2 = \textbf{15}\_$$

Step Two: 4 x 2 = 8

$$\textbf{34} \times 5\ \textbf{2} = 15\ \_\textbf{8}$$

Step Three: RAINBOW!

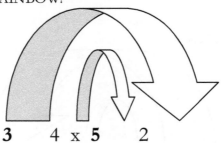

$$3 \qquad 4 \times 5 \qquad 2$$

$$
\begin{array}{rr}
3 \times 2 = & 6 \\
4 \times 5 = & +2\,0 \\
\hline
& 2\ \underline{\textbf{6}}
\end{array}
$$

Put the number '6' on top of the underline and carry the '2'

$$34 \times 52 = 15\ \underline{6}\ 8$$
$$\textbf{2}$$

Now add 15 + 2 =17

So the final answer:
$$34 \times 5\,2 = 17\ \underline{6}\ 8$$

Example Four: 86 x 5 7

$$8\,6 \times 5\,7$$

Step One: 8 x 5 = 40
$$8\,6 \times 5\,7 = \mathbf{40\_}$$

Step Two: 6 x 7 = 4 **2**
$$86 \times 5\,7 = 40\ \_\mathbf{24}$$

Step Three: RAINBOW!

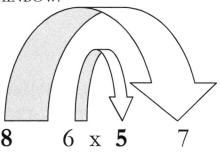

8   6 x **5**   7

$$8 \times 7 = \quad 5\ 6$$
$$6 \times 5 = \underline{+ 3\ 0}$$
$$_8\ \mathbf{6 +4 =90}$$

Put the number '6' on top of the

underline and carry the '2'

$$86 \times 5\,7 = 40 \; \underline{0} \; 2$$
$$\mathbf{9}$$

Now add 40 + 9 = 49

So the final answer :
$$86 \times 5\,7 = 49 \; \underline{0} \; 2$$

Once you have done a few of these, then you will see how quickly to can do the multiplications using the Rainbow method.

Exercise (RB1)
(You can check these using a calculator once you have completed each question !

There is a famous saying:

"FEEDBACK IS THE BREAKFAST
OF CHAMPIONS"

1. 12 x 15
2. 23 x 22
3. 26 x 27
4. 31 x 33
5. 42 x 45

6. 53 x 54
7. 61 x 63
8. 72 x73
9. 81 x 83
10. 93 x 91

Here is a few more challenging exercise for you Child Genius!

Exercise (RB2)

1. 27 x 36
2. 38 x 49
3. 67 x 56
4. 83 x 76
5. 77 x 58
6. 84 x 96
7. 58 x 78
8. 88 x 69
9. 71 x 98
10. 75 x 89

With practice, you will soon be able to do these mentally and quickly!

Don't be surprise when other people will start asking you (the Child Genius):

"HOW DO YOU DO THAT" ?
Your answer: It's obvious!

Answers to the exercises:

RB1

1. 180
2. 560
3. 702
4. 1023
5. 1890
6. 2862
7. 3843
8. 5256
9. 6723
10. 8463

RB2

1. 972
2. 1862
3. 3752
4. 6308
5. 4466
6. 8064
7. 4524
8. 6072
9. 6958
10. 6675

Personal Notes

CHAPTER NINE

WORKING OUT ANY 2 DIGIT SQUARE NUMBERS

(USING THE RAINBOW

MULTIPLCATIONS METHOD)

Once you have mastered the basic skills of the Rainbow method, you can use this to work out any square numbers from 10 to 99.

This is particularly useful for those of you doing tests and examinations, like students in the UK doing their GCSE Mathematics for the Non-Calculator paper.

Example One: 13^2

$$\mathbf{1}3 \times \mathbf{1}3$$

Step One: 1 x 1 = 1

$$\mathbf{1}3 \times \mathbf{1}3 = \mathbf{1}\_$$

Step Two: 3 x 3 = 9

$$1\,\mathbf{3} \times 1\,\mathbf{3} = 1\,\_\mathbf{9}$$

Step Three: RAINBOW!

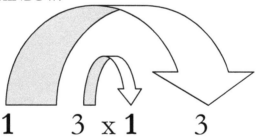

$$1 \quad 3 \times \mathbf{1} \quad 3$$

$$
\begin{array}{r}
1 \times 3 = \quad 3 \\
3 \times 1 = \underline{+3} \\
6
\end{array}
$$

Therefore, 13x13 = 169

Example Two: 34^2

$$3 4 \times 3 4$$

Step One: $3 \times 3 = 9$

$$3 4 \times 3 4 = 9\_$$

Step Two: $4 \times 4 = 16$

$$3 \mathbf{4} \times 3 \mathbf{4} = 9\_6$$
$$1$$

Step Three: RAINBOW!

$$\mathbf{3} \quad 4 \times \mathbf{3} \quad 4$$

$3 \times 4 = \quad 12$
$4 \times 3 = \quad \underline{+ 12}$
$ 24$

And add the 1, $24 + 1 = 25$

$$34 \times 34 = 9 \, \underline{56} \qquad (9 + 2 = 11)$$
$$2$$

So the final answer is 11 56

Example Three: 34^2

$$\mathbf{8}\,9\,\mathrm{x}\,\mathbf{8}\,9$$

Step One: 8 x 8 = 64

$$\mathbf{8}\,9\,\mathrm{x}\,\mathbf{8}\,9 = \mathbf{64}\,\_\qquad (8x8=64)$$

Step Two: 9 x 9 = 81

$$8\,\mathbf{9}\,\mathrm{x}\,8\,\mathbf{9} = 64\,\_\,\mathbf{1}$$
$$\qquad\qquad\qquad\;_{8}$$

Step Three: RAINBOW!

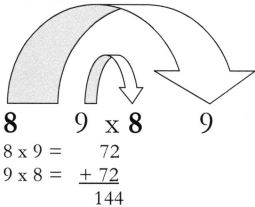

$$\mathbf{8}\qquad 9\;\mathrm{x}\;\mathbf{8}\qquad 9$$

8 x 9 =   72
9 x 8 =   + 72
                144

And add the 8, 144 +8 = 152

$$89x89 = 64\,\underline{2}1 \qquad (64+15=79)$$
$$\qquad\qquad\quad_{1\,5}$$

So the final answer is 79 21

Personal Notes

CHAPTER TEN

THE 10-SECONDS CHILD GENIUS

MY JOURNEY SO FAR (TRAY-SEAN BEN SALMI)

THERE IS NO SUCH THING AS 'FAILURE' ONLY FEEDBACK

I was once a very shy person(still am!) and I would not speak to anyone **and I do mean anyone!**
I was always afraid of what others thought about me.
My mum started taking my sister and I to learn loads of things and we started to speak our mind and bit by bit we started speaking on stage and we used to make our family anthem.

THE JOURNEY OF A THOUSAND MILES BEGIN WITH ONE STEP AT A TIME.

After becoming more confident and
<u>learning to ask not only more questions but the right ones</u>!
I found an artist to design the logo for my Brand and started to
make T-Shirts and came up with my brand called:

I'M THAT KID

and as the brand continued to expand not only did I sell T-Shirts
I also started to sell wristbands and running events
Yes, Me – that shy quiet kid who was afraid to talk to anyone!

THERE IS NO SUCH THING AS A SELF-MADE MAN

After I learnt from all these amazing people they have completely changed my mindset and the way I think in regards to money.

I will never forget what some of them said to me

YOU HAVE TO GET
COMFORTABLE
WITH
BEING
UNCOMFORTABLE
BEFORE
IT GETS
COMFORTABLE

Personal Notes

THE AUTHOR'S CHARITY

We will donate £1 of every book to go towards
I'M THAT KID FOUNDATIONS
-    Paying It Forward which we want one day to
fund someone else to write their book
To inspire others.

THE AUTHORS' OTHER

CHOSEN CHARITY

IS

UNICEF

Disclaimer:-

THE AUTHORS' CHOSEN CHARITY TO FUND RAISE
FOR 'Unicef UK' - We are individuals taking independent
actions to fundraise vital money for Unicef's important work
for children. We are NOT endorsed by nor work for Unicef
UK. We will donate at least £1 of the proceeds from our
book to Unicef UK.

Author's Recommendations

BOOKS:

- How To Be A Student Entrepreneur by Junior Ogunyemi
- Rich Dad, Poor Dad for Teens by Robert Kiyosaki
- Conversations with God for teens By Neale Donald Walsch
- The Little Soul and The Sun by Neale Donald Walsch
- The Little Soul and The Earth by Neale Donald Walsch
- Raising CEO Kids by Dr Jerry Cook & Sarah Cook
- Who moved my cheese by Dr Spencer Johnson
- Think and Grow Rich by Napoleon Hill
- Key Person of Influence by Daniel Precisely
- The Secret by Rhonda Byrne
- The Hidden Messages in Water by Dr Masaru Emoto
- I THINK I AM by Louise L. Hay
- The Story of The Ethical Elephants by Catherine Warrington

DVDs:

- The Secret
- What The Bleep Do You Know?
- Thrive

Personal Notes

ABOUT THE AUTHORS

TRAY-SEAN BEN SALMI

"You're too young"

As Tray-Sean is writing this book, he has just turned 12 years old
and after lots and lots of challenges,
Today his Brand is called "I'm That Kid"

His aim is to inspire young children from whatever background to
celebrate their uniqueness and go after their Dreams.

Together with his four other siblings:

LASHAI, PAOLO, YASMINE AND AMIRE.
They are known as

"The Fantastic Five"

**Their biggest mentor and motivator
their mum
SABRINA BEN SALMI.**

Here is just a sample of his achievements so far.

Tray-Séan Ben Salmi (12yrs old) -
https://www.gofundme.com/2mj97zyk

He is the co-author of Kidz That Dream Big, UnLtd
Award Winner,
Nominated for a R.E.E.B.A Award 2017,
Child Genius 2017 final 20, Nominated for National
Diversity Award 2017,
Radio show host, Winner of Radio Works Authors
Awards 2017,
Co-founder of Put The RED CARD up to bullying
and founder of I'M THAT KID - Bridging The Gap
Between Fathers & Sons.
I'm That KID is also a t-shirt brand - teaching boys
to simply shine their unique light of positivity (I'm
That KID covers:

I'm That KID – Creating A Vision Board For My
Future,
I'm That KID – Taking The Stage, I'm That KID -
Inspiring My Community To Pay It Forward, I'm
That KID - There's A Inside ME and I'm That KID
- Families That Play Together, Stay Together).
Harry Singha Foundation - YLS (Young Leaders
Summit) Fellow,
Access Consciousness Bars Practitioner and NLP
Practitioner.
Tray-Séan launched his first short story competition
in school at the age of 6 and has been attending a

variety of investment, business and personal development events such as Andy Harrington Power to Achieve,
How To Build A Brand, T Harv Eker MMI and TEDx Salford.
Tray-Sean was invited to attend the TEDx Salford speakers dinner with the 12 speakers one of them being a NASA astronaut.

He went from being an extremely shy boy to becoming a certified public speaker, author, coach and has been doing work experience at a branding company for almost two years in March.
The company he works with is called How to Build A Brand and he tweets for them like crazy.
He is also a passionate social entrepreneur, youth coach and a budding artist and fashion designer.
He's currently in the process of launching his smart clothing brand "Dapper by Tray-Séan Ben Salmi" the range will start off with Cuff links, a flat cap, ties and bow ties and there's a finishing attached to prepare young boys for manhood.
Tray-Séan is on the gifted and talented list at school and above national average on all subjects.

One day he dreams of becoming a mathematician.
Tray has a tutor called Ras Al, and due to being academically advance, he's on target to sit his GCSEs early.

The inspiration behind Kidz That Dream Big! was to end, once and for all, what Lashai & Tray-Séan saw as "the dubious concept of so called public school education which missed financial education, emotional intelligence and personal development". Tray-Séan feels that financial education, Emotional Intelligence and Personal Development are too important to missed.
Together with the Kidz That Dream Big book, coaching and workshops children and youth can seed their time and financial freedom future today.

He has co-authored a book with his older sister
Lashai called
"Kidz That Dream Big"
- Essential Tips on How to Make Money Doing
What You Love –

Once of his big initial achievement was conducting a workshop bridging the gaps between Fathers and their son(s) and help them to communicate more effectively and lovingly to each other.
This was held at the famous venue of the top British professional football club ARSENAL FC.

He has held other workshops called
"I'm That Kid – Dreamboard"
where children are let loose with their imaginations, inspiring all the kids to dream big.

"Remember, every modern invention we have today was ONCE just a Dream in someone's imagination"

Tray-Sean is a Coach and Mentor to help young people
(and some adults) to become more confident
and his training is done with his clients via Skype.

It would be his pleasure to one day to reach a very wide audience,
including yourself at his webinars,
live workshops and other events including his Retreats at some of
the Exotic locations in the World.

Would you like to join him and have fun in wonderful warm Exotic
locations in the warm sun and beautiful beaches?

Philip Chan
10 Seconds Maths Expert
Award Winning Author
International Award Winning Radio Host

www.10secondstochildgenius.co.uk

PHILIP CHAN

Philip Chan has been a successful teacher for over forty years and he has extensive experience in teaching at all levels of expertise. This ranges from Primary School level through to High Schools, plus College Students, as well as Adult Education.

Philip has conducted hundreds of mathematics workshops working with children and their parents together. Over a number of years using his vast skills mentoring trainee teachers to empower them in the classroom and workshops using his fun and unique techniques, to create excitement and confidence in an instant.

Many of his students have successfully gone on to some of the leading universities such as Oxford and Cambridge to gain their PhD and First Class Honors Degrees, as well as a number of his student becoming top leaders and appointed to senior positions in both business and education.

Philip is a qualified Life Coach and NLP Practitioner working with groups and individuals on personal development. Philip is a former Elite Sports Performance Coach and has helped many athletes progress to competing at National, International and Olympic standards.

He has been successfully working and mentoring some of the top Executives from UK Blue Chip companies and helped several Global Billion Dollar companies with their expansion plans and development over a number of years.

For more than forty years, Philip has been involved in fund raising for a number of charities, including UNICEF, Shelter, Oxfam, YMCA and many others by giving informative talks on subjects like: Stress Management; Prevention and Recovery from serious illnesses, such as cancer, without the use of drugs. Other talks include Relaxation Techniques and Memory Training in preparation for academic examinations. All donations go directly to the charities concerned.

Philip is also a motivational speaker and has enriched the lives of countless people in achieving their goals and dreams. Currently he is working with a number of the world's top business coaches from the UK, Canada, USA, South Africa and Australia plus other countries to develop businesses for mutual benefit.

He is a Double Amazing Best Selling author, one of the few authors in the world to have two books going to Number One in two consecutive months in UK and Australia simultaneously.

He is an International Winning Award Radio Hosts in over 70 countries in the world.

His goal and dream is to inspire the next generations to develop a positive attitude for learning in all subjects and empower them in the joy of learning, discovery and raising their self beliefs for greater achievement.

His goal is to share these techniques with children and empower teachers across the world them with his '10-Seconds Speed Maths Techniques' and Founder of Jet Set Learning.

He would love you to be a part of this success to SHARE THE KNOWLEDGE.

Raymond Aaron

NY Times No.1 Bestselling Author

THERE IS NO SUCH THING AS A DISABLED PERSON

ONLY

THIS-ABLE PERSON!

Famous People with Learning Disorders

Did you know that many of famous celebrities have struggled with basic learning skills growing up? And yet despite the ridicule and bullying at school, they often suffered in school, by their so called 'Friends' and even family, they had the inner strength to persevere and overcome.

Here are a few such people!

Keira Knightley (Dyslexia)

Diagnosed with dyslexia at age 6, famous actress Keira Knightley has said her struggles with reading at an early age only made her tougher.

There is no such thing as a 'Disabled' person – only a THIS-able person

Orlando Bloom (Dyslexia)

Orlando was best known for his role as Will Turner in Pirates of the Caribbean. He was diagnosed with dyslexia at age 7 and his struggles left him looking for a creative outlet, so he turned to the stage. He called his Dyslexia as a 'GIFT' because he had to learn everything forward and backward, inside out, so he was fully prepared.

There is no such thing as a 'Disabled' person – only a THIS-able person

Michael Phelps (ADHD)

This most highly decorated Olympian of all time, boasting 22 medals (18 of them being gold) was continually criticized by teachers for his inability to sit still, and was formally diagnosed with ADHD when he was in fifth grade. After being on Ritalin for over two years, Phelps chose to stop using the drug and instead used swimming to help him find focus.

There is no such thing as a 'Disabled' person – only a THIS-able person

Daniel Radcliffe (Dyspraxia)

Famous for his role as Harry Potter, Daniel Radcliffe has lived with a mild case of dyspraxia for his entire life.

He still has trouble tying his shoelaces.

But determinations and hard work wins the day!

There is no such thing as a 'Disabled' person – only a THIS-able person

Whoopi Goldberg (Dyslexia)

Famous for leading roles in movies like Sister Act and other films, now she is not only an actress, writer, and producer as well. Whoopi was actually called "dumb" while growing up due to her dyslexia.

But she understands her self worth and the rest is history as they say!

There is no such thing as a 'Disabled' person – only a THIS-able person

Steven Spielberg (Dyslexia)

This legendary filmmaker of: Indiana Jones, E.T. Saving Private Ryan, and Jurassic Park and many more films.

He was only diagnosed with dyslexia at age 60 and he had struggled with it his entire life. He bullied so much that he dreaded going to school.

He offers this advice to students and young adults with learning disabilities, "You are not alone, and while you will have dyslexia for the rest of your life, you can dart between the raindrops to get where you want to go. It will not hold you back."

There is no such thing as a 'Disabled' person – only a THIS-able person

Justin Timberlake (ADD and OCD)

Singer, songwriter, and actor .Justin Timberlake revealed that he has both Attention Deficit Disorder and Obsessive Compulsive Disorder and is quoted as saying "I have OCD mixed with ADD.

Despite battling his obsessive thoughts, he has had an incredibly successful career in the entertainment industry, even winning nine Grammy Awards and four Emmy Awards so far!

There is no such thing as a 'Disabled' person – only a THIS-able person

Jamie Oliver (Dyslexia)

The famous celebrity Jamie Oliver has now written over twenty cookbooks, and currently holds the title of world's richest chef, with a net worth of over $230 million.

He was once quoted as saying "I've never read a book in my life.

There is no such thing as a 'Disabled' person – only a THIS-able person

Keanu Reeves (Dyslexia)

The film star of many movies like The Matrix trilogy, Point Break, Bill and Ted's Excellent Adventure, and many more films.

He struggled badly at school with his dyslexia.

He once said, "Because I had trouble reading, I wasn't a good student ... I didn't finish high school. I did a lot of pretending as a child. It was my way of coping with the fact that I didn't really feel like I fit in."

His gift for pretending has served him well in his acting career, which is still going strong after 30 years.

There is no such thing as a 'Disabled' person – only a THIS-able person

Charles Schwab (Dyslexia)

At 77 years old, Businessman and investor Charles Schwab has a net worth in excess of $5.1 billion, and yet still finds reading and writing tedious.

Charles Schwab bluffed his way through his early years of schooling. Due to his struggle with undiagnosed dyslexia for many a years.

There is no such thing as a 'Disabled' person – only a THIS-able person

Jay Leno (Dyslexia)

A former television host of NBC's The Tonight Show. He is also a comedian, actor, writer, producer, voice actor.

His dyslexia has led him to become a firm believer in low self-esteem, in that "If you don't think you're the smartest person in the room and you think you're going to have to work a little harder, and put a little more time into it to get what everybody else does, you can actually do quite well.

That's been my approach." His approach to dyslexia has clearly paid off.

There is no such thing as a 'Disabled' person – only a THIS-able person

Henry Winkler (Dyslexia)

Best known as Arthur Fonzarelli, aka "The Fonz" in the classic television show Happy Days.

Since his diagnosis at age 31, Winkler has become a champion for those suffering from dyslexia, and has even authored a series of books about a child with dyslexia that is based upon his own experiences with the disorder, Hank Zipzer: The World's Greatest Underachiever.

There is no such thing as a 'Disabled' person – only a THIS-able person

Cher (Dyslexia)

World famous for her singing and as an actress, not many people are aware that Cher has had problems with dyslexia her entire life.

Her dyslexia went undiagnosed in school; her teachers simply thought that she wasn't trying.

In an interview she said: "When I was in school, it was really difficult. Almost everything I learned, I had to learn by listening. My report cards always said that I was not living up to my potential." She also admitted that dyslexia had made it more difficult to read movie scripts, but that though it slowed her down she refused to let it stop her.

There is no such thing as a 'Disabled' person – only a THIS-able person

Richard Branson (Dyslexia)

World famous Entrepreneur and billionaire for his famous VIRGIN Brand.

He is "The only person in the world to have built eight billion-dollar companies from scratch in eight different countries." Richard Branson is a model for success, he is also dyslexic.

Unlike many, who consider dyslexia a curse, Branson calls it his "greatest strength

His former headmaster who said, "Congratulations, Branson. I predict that you will either go to prison or become a millionaire." Looking back on the incident Branson said "That was quite a startling prediction, but in some respects he was right on both counts!"

There is no such thing as a 'Disabled' person – only a THIS-able person

Personal Notes

SPECIAL

PEOPLE

This is one of our very Special and Amazing friend

DREW HUNTHAUSEN

Also known as:

The No Excuses Blind Guy

This is our amazing friend, Drew **Hunthausen** and his brief Bio below:

An International Speaker and Bestselling Author.

He went from an 11 Year old playing baseball and golf, to **suffering near death** in a 3 month coma brought on by bacterial meningitis.

He emerged from the coma **totally blind, hearing impaired,** and unable to even sit-up by himself. It took him 7 long years of physical therapy to reach a fraction of where he is today.

Today Drew has resumed Snow Skiing, is an awarded Triathlete, and inspires audiences to live an extraordinary life through his message of meeting life's challenges head on with love, gratitude, encouragement, and no excuses!

For his fully story, you can contact him below.

Drew Hunthausen

The No Excuses Blind Guy

Drew@NoExcusesBlindGuy.com

www.DrewsInspirations.com

www.BookDrew.com

Personal Notes

The eight things parents say about maths at their child(rens) school are:-

1. They do it so different to when I was at school.

2. My kids get annoyed with me when I try to help.

3. The teachers don't know how to explain it in a simple way and I am clueless about algebra and their teachers too.

4. My kids are so confused about Fractions.

5. I don't know how to help my kids with their maths because I was unless at school in this subject.

6. My kids keeps forgetting all the times tables.

7 The fact that I can't help them as they've been taught very differently to when I was at school.

8 Boring! Lots of learning numbers with nothing interesting to relate it to....

What are my needs for my Children?

Personal Notes:

www.**10secondstochildgenius**.co.uk

Did you know this about your Authors?

TRAY-SEAN BEN SALMI

The S.H.Y. Kid

S :   **SMILING**

H :   **HAPPY**

Y :   **YOUTH**

Now, "I'M That Kid"

PHILIP CHAN

I was bottom of the class at school and medically 'Disabled'

Now, "The 10-Seconds Maths Expert" and

THIS-able Person.

It is not where you are NOW that is important BUT where you want to be by

TAKING THE RIGHT ACTIONS

NOW

Embrace your GENIUS WITHIN!

YOUR DECLARATION TO YOURSELF

I

_____

(INSERT YOUR NAME)

Promise to always be open minded to seek out to be the Best Person I can be.

Eventually, I will achieve all the OUTCOMES I want because I will KEEP ON KEEPING ON UNTIL I SUCCEED.

Signed: _____

Date: _____

My Goals:

(Also put a date when you want to achieve these goals)

SHORT TERM

1.

2.

3.

4.

5.

MEDIUM TERM

1.

2.

3.

4.

5.

SHORT TERM

1.

2.

3.

4.

5.

We (Tray-Sean and Philip) want to wish you all the success to bring out the CHILD GENIUS within you.

QUESTION:

How do you know what you don't know?

ANSWER:

When someone can see your potentials and tell you!

With love and blessings

On your journey.

Tray-Sean Ben Salmi – I'm That Kid

and

Philip Chan – The 10 Seconds Maths Expert

www.10secondstochildgenius.co.uk

Personal Notes:

We have launched our first 10 Seconds To Child Genius event at the London Marriot West India Quay in 2017.

This was the first of the many 10 Seconds To Child Genius Workshops to come.

Workshop One:
Discover the Genius Within.
Welcome to the 10 Seconds To Child Genius Academy.

Workshop Two:
Working towards Masterclass.

Workshop Three:
Doing the Impossible!

To keep up to date, please visit:
www.10secondstochildgenius.co.uk

Lightning Source UK Ltd.
Milton Keynes UK
UKHW021911220419
341424UK00021B/485/P